WHITMAN MIDDLE SCHOOL LIBRARY

WHERE DO WE LOOK FOR LIFE?

by Dru Hunter

CREATIVE EDUCATION • CREATIVE PAPERBACKS

Published by **Creative Education** and **Creative Paperbacks**
P.O. Box 227, Mankato, Minnesota 56002
Creative Education and Creative Paperbacks are imprints of The Creative Company
www.thecreativecompany.us

Design and production by **Christine Vanderbeek**
Art direction by **Rita Marshall**
Printed in Malaysia

Photographs by Alamy (BRUCE COLEMAN INC., Steve Bloom Images), Corbis (Bettmann, Antar Dayal/Illustration Works, Kyu Furumi/Aflo, Kam & Co./cultura, Karen Kasmauski, Frans Lanting, Lebrecht Music & Arts/Lebrecht Music & Arts, Alfredo Dagli Orti/The Art Archive, Cheryl Power/Science Photo Library, Science Photo Library, Paul D.Stewart/Science Photo Library, Thomas Kitchin & Victoria Hurst/All Canada Photos, Underwood & Underwood), Dreamstime (Alle, John Anderson, Outdoorsman), iStockphoto (imgendesign, John Pitcher), Shutterstock (Danny Alvarez, Norma Cornes, dencg, f9photos, Oleg Golovnev, Brendan Howard, Ian Kennedy, Georgios Kollidas, Ungnoi Lookjeab, Dudarev Mikhail, Stefan Pircher, Jose AS Reyes, Severe, Marina Sun), SuperStock (Minden Pictures), Wikipedia (Jens Petersen)

Copyright © 2016 Creative Education, Creative Paperbacks
International copyright reserved in all countries. No part of this book may be reproduced in any form without written permission from the publisher.

Library of Congress Cataloging-in-Publication Data
Hunter, Dru.
Where do we look for life? / Dru Hunter.
p. cm. — (Think like a scientist)
Includes bibliographical references and index.
Summary: A narration of the origins, advancements, and future of the life sciences, including botany and zoology, and the ways in which scientists utilize the scientific method to explore questions.

ISBN 978-1-60818-596-2 (hardcover)
ISBN 978-1-62832-201-9 (pbk)
1. Life sciences—Juvenile literature. 2. Science—Methodology—Juvenile literature. 3. Scientists—Juvenile literature. I. Title.

QH309.2.H86 2015
570—dc23   2014033141

CCSS: RI.5.1, 2, 3, 8; RI.6.1, 3, 7; RST.6-8.1, 2, 5, 6, 8

First Edition HC 9 8 7 6 5 4 3 2 1
First Edition PBK 9 8 7 6 5 4 3 2 1

ON THE COVER A giant tortoise from the Galápagos Islands near Ecuador

WHERE DO WE LOOK FOR LIFE?
TABLE OF CONTENTS

INTRODUCTION 5

CHAPTERS

LIFE SCIENCE EVOLVES 7
SURVIVAL OF THE FITTEST 17
IN THE FIELD 27
YEAR OF THE HORSE 37

GLOSSARY 46
SELECTED BIBLIOGRAPHY 47
WEBSITES 47
INDEX 48

SCIENTIST IN THE SPOTLIGHT

EDWARD JENNER 10
LOUIS PASTEUR 20
RACHEL CARSON 30
ELIZABETH BLACKWELL 40

WHERE DO WE LOOK FOR LIFE?

INTRODUCTION

Peering through his handcrafted microscope, Anton van Leeuwenhoek examined the bee under the lens. He could see the insect had tiny mouthparts and a needlelike stinger. He reported his first microscopic observations of the bee, a fungus, and a louse to the Royal Society of London, but many doubted his work until they could replicate his experiments.

Leeuwenhoek continued building microscopes—more than 500 of them—in order to look more closely at substances such as lake water, blood, and even tooth scrapings. In 1676, he announced that he had observed tiny organisms moving around inside drops of water. When another scientist was able to confirm those results, it was determined that Leeuwenhoek was the first person to observe bacteria!

From the smallest bacterium to the largest whale, the life sciences encounter and study life in all its varied forms. Life scientists use the **scientific method** to gather knowledge, investigate, observe, and experiment to learn about the world. Studying life sciences can help solve human health problems, improve conservation methods, and foster coexistence among Earth's living things. Whether it was Leeuwenhoek observing human blood cells for the first time, botanists finding that plant leaves undergo a process called **photosynthesis**, or marine biologists locating the elusive giant squid in its deep-ocean habitat, amazing discoveries are made possible through the life sciences.

WHERE DO WE LOOK FOR LIFE? CHAPTER 1

LIFE SCIENCE EVOLVES

LIFE SCIENCE IS THE STUDY OF ALL LIVING ORGANISMS, from Earth's smallest **microorganisms** to its largest animals, tallest plants, and biggest **super-organisms**. Life scientists might study cells, insects, humans, blue whales, giant sequoia trees, or the Great Barrier Reef. By understanding how organisms live, survive, and reproduce, life scientists can improve not only human lives but the lives of other living things as well. The main science behind the life sciences is biology. Over the years, many branches of life science have formed, including botany (the study of plants), microbiology (the study of microorganisms), genetics (the study of **heredity**), and zoology (the study of animals).

Biology developed into its modern form through contributions from ancient civilizations. Early people observed animals and plants but didn't always realize the connections of living things to their everyday lives. The ancient Egyptians often explained disease as an evil demon and described

Zoologists analyze growth and behavior in animals such as humpback whales.

7

LIFE SCIENCE EVOLVES

a bountiful harvest as the work of gods who were pleased. However, they and other ancient peoples from China, the Middle East, and India developed knowledge of plants that helped with ailments. Plants such as dried lichen were even found packed in the abdomens of Egyptian mummies, and oils made from juniper berries played a role in embalming (preserving) bodies during mummification.

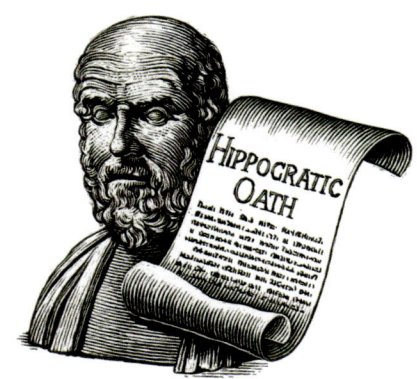

Whether celebrated in India (opposite) or studied for medicine in Greece (below), plants have worldwide value.

Perhaps no civilization influenced modern biology more than the ancient Greeks. Hippocrates of Cos is believed to have been born around 460 B.C. He was the first to study disease and use reason, or logic, to explain the causes for failing health. Hippocrates's methods from thousands of years ago resemble a doctor's clinical observations today. He wrote many books on procedures that doctors should use when examining a patient. In his book *On Forecasting Diseases*, he writes, "First of all, the doctor should look at the patient's face. If he looks his usual self, this is a good sign. If not, however, the following are bad signs—sharp nose, hollow eyes, cold ears, dry skin on the forehead, strange face color.... If the face is like this at the beginning of the illness, the doctor must ask the patient if he has lost sleep, or had diarrhea, or not eaten." Often referred to as the "Father of Medicine," Hippocrates is remembered for his **ethical** beliefs about the practice. Today, new graduating doctors from medical schools recite the "Hippocratic Oath," a promise to take responsible and discreet care of patients, forbidding any harm to them.

Aristotle, who lived in Greece from 384 to 322 B.C., also contributed to biology. He considered the study of animals to be central to the understanding of nature. Because Aristotle was able to organize his scientific inquiries, he wrote the equivalent of encyclopedias on the different phases of animal life processes such as birth. In book two of his *Posterior Analytics*, he explains, "The things about which

SCIENTIST IN THE SPOTLIGHT
EDWARD JENNER

From childhood, Edward Jenner (1749–1823) had an interest in many areas of science. When he was 13, he worked as an apprentice to a surgeon in England. After becoming a physician, he conducted many experiments with human blood. During his years of treating illnesses, he observed that dairymaids who had been exposed to the cowpox disease didn't contract the often-fatal smallpox. He **hypothesized** that cowpox offered immunity, or resistance, to smallpox and began experimenting with ways to use that information to protect others. He took a sample from a dairymaid's fresh cowpox outbreak and **inoculated** an eight-year-old boy. After the boy was exposed to smallpox, he did not catch the disease. Jenner is often referred to as the "Father of Immunology" because his science spurred a new field in biology. Vaccinations led to the worldwide eradication of smallpox in 1980, saving countless lives.

LIFE SCIENCE EVOLVES

we inquire are equal in number to the things we understand. We inquire about four things: the fact, the reason why, if something is, what something is."

Galen was a Greek who lived in Rome around A.D. 130 to 201. A plague that struck in 166 has been called the Plague of Galen because of his study and treatment of the disease known today as smallpox. Galen also made detailed notes on the anatomy of animals, which were frequently used during the Middle Ages by those studying medicine.

Some of the hypotheses, data, and conclusions of the ancient life scientists were correct. However, inaccurate information often circulated until debunked by more advanced science. One such false belief was that the human body consisted of four liquids: snot, blood, black bile, and yellow bile. Another was the widely held acceptance of spontaneous generation. This was the theory that living organisms could be born from nonliving or dead matter.

Leeuwenhoek's many observations also contributed to the study of reproduction in plants and animals.

Biology took a giant leap forward in the 1590s with the invention of the compound microscope. This instrument used light and a system of lenses to magnify objects to nine times their original size, but the images were a little blurry. Then, in the mid-1600s, Leeuwenhoek made improvements to the microscope, grinding and heating small glass lenses until they had a curvature that could magnify objects up to 270 times their original size. This led to many new discoveries, such as bacteria and infusoria (tiny water creatures). By utilizing the microscope, Jan Swammerdam (1637–80) of Amsterdam, Netherlands, advanced the field of entomology (the study of insects). Swammerdam had a medical degree but spent much of his life researching thousands of insects. He came up with special techniques to view the insects using

DID YOU KNOW? One square inch (6.5 sq cm) of human skin contains as many as 20 million organisms.

Biology examines all life forms, from human skin cells (left) to butterfly transformation (above).

microscopes. With his meticulous drawings, he detailed the stages of butterfly **metamorphosis**. Swammerdam also became well known for discovering valves in the human lymphatic system, which were later named after him. His work led to the identification of red blood cells. They carry oxygen from the lungs to other cells in the body.

In the 1830s, the Germans Theodor Schwann and Matthias Schleiden changed how life scientists thought about living cells. They believed that cells were the basic units and building blocks of living things, whether plants or animals. Schwann and Schleiden's early cell theories became the cornerstone of modern biological understanding and research.

Biology and its branches organize living organisms by taxonomy, a system that classifies each organism by its structure and function. Swede Carl Linnaeus (1707–78) wanted to make a system that was easy for other scientists to remember and use. In his book *Systema Naturae*, he introduced five classification categories, beginning with the three kingdoms of animal, vegetable (plant), and mineral. The way Linnaeus organized the plants by number of stamens (the stemmed, pollen-making organs) revolutionized the classification of those organisms. Further breaking down the natural world into class, order, genus, and species, Linnaeus's system grouped humans together with monkeys—as members of the primate order—for the first time. In later editions of his book, Linnaeus made some adjustments, such as switching whales from the fish to the mammal class.

While teaching at Uppsala University, Linnaeus had his students go on explorations by sea and land. One student named Daniel Solander was on captain James Cook's first around-the-world expedition and returned with plants from Australia and the South Pacific. Linnaeus's classification of life eventually grew from a small pamphlet to volumes of books as people sent plant and animal

LIFE SCIENCE EVOLVES

specimens to him from all over the world, asking for scientific descriptions. Some of his other interests involved trying to discover native substitutes for commercial crops not found in Sweden. But because of Sweden's cold weather, he was unsuccessful at growing bananas, rice, coffee, tea, and cacao in his homeland's soil.

Taxonomy is important because it helps scientists understand an organism's **evolution**. Classifying assists in conservation because it helps scientists identify what kinds of plants or animals have died out in the past and how diverse or limited some living things are at present. Classification of organisms is also useful in medicine, such as when scientists study whether a disease can spread between species. For more than 200 years, Linnaeus's classification system has been studied and modified by life scientists—including one of biology's most famous naturalists, Charles Darwin.

Thanks to Linnaeus (above), each species has a two-word scientific name, such as Plusiotis optima (opposite).

TRY IT OUT! Gather your friends, a thermometer, a notebook, and a stopwatch. Record your friends' temperature before and after giving them a timed multiplication test. Analyze your results: Did the stress affect their body temperature? What other factors might have played a role?

WHERE DO WE LOOK FOR LIFE? CHAPTER 2

SURVIVAL OF THE FITTEST

IN 1831, THE HMS *BEAGLE* SET SAIL FROM ENGLAND ON A five-year exploration. On board the ship was 22-year-old naturalist and geologist Charles Darwin. His previous research included studies of marine **invertebrates**, black spores, and beetles. The *Beagle* expedition would give him the opportunity for hands-on research, observation, and experiments in botany and zoology. Traveling around the Pacific Islands, South America, and the Galápagos Islands, Darwin collected **fossils** and more than 1,500 bottled specimens of birds and plants. While contemplating the formation of atolls (ring-shaped islands made of coral), he laid out what are now the foundations of modern coral reef theories.

Darwin was endlessly fascinated by the nature he observed. After coming ashore in South America, he wrote, "The noise from the insects is so loud that it may be heard even in a

The HMS Beagle's *legendary journey helped change the course of science.*

17

SURVIVAL OF THE FITTEST

vessel anchored several hundred yards from the shore; yet within the recesses of the forest, a universal silence appears to reign. To a person fond of natural history, such a day as this brings with it a deeper pleasure than he ever can hope again to experience."

In 1835, Darwin arrived in the Galápagos Islands, located about 600 miles (966 km) west of Ecuador. This five-week stop during the *Beagle*'s voyage would become one of the most famous events in the history of science. Previously, Darwin and other scientists had accepted certain theories about nature—but his experiences in the Galápagos convinced him that there was more to the biological world than they had thought.

The Galápagos Islands were home to numerous unique species of plants, reptiles, and birds. These species had developed in complete isolation from mainland and other island species. Darwin concluded that these plants and animals had transformed over time and developed different characteristics. He studied and collected many specimens, including finches that were later named after him. He determined that the finches had arrived on the island as one form. As years went by, however, the finches flew to other islands and developed into the 13 different kinds of finches Darwin encountered, each one **adapted** for survival in its new home.

Darwin also studied the islands' giant tortoises. With shell lengths of more than 5 feet (1.5 m) and weights of more than 500 pounds (227 kg), the tortoises differed in appearance from island to island. The finches and tortoises gave Darwin evidence to support his theory of **natural selection**.

When Darwin returned to England from his life-changing voyage, he brought with him discoveries that would also change biology. Darwin mulled over his hypotheses, finally publishing his theory

Darwin noted variations among the Galápagos tortoises' shells (opposite) and the finches' beaks (below).

19

SCIENTIST IN THE SPOTLIGHT
LOUIS PASTEUR

Louis Pasteur (1822–95) was born in France. After earning his doctorate, he became a chemistry professor and used experiments to show that bacteria caused milk to go sour. He went on to invent a process, known as pasteurization, still in use today in which bacteria is destroyed by first boiling and then cooling food. Despite becoming partially paralyzed from a stroke in 1868, Pasteur continued his microbe research. When silkworm disease threatened the silk industry, he discovered that, by getting rid of the parasitic organisms in silkworm eggs, the disease would disappear. In 1879, Pasteur made an important germ discovery by accident. When he exposed chickens to a virus **culture**, he noticed that the chickens became immune to the virus. This realization led to the vaccines for cholera, tuberculosis, and smallpox. After a nine-year-old boy was bitten by a dog with rabies, Pasteur successfully vaccinated him, too.

of evolution and explanations of natural selection in 1859's *On the Origin of Species*.

According to Darwin, natural selection means that in all living populations, there are variations among the same species. There will be individuals born with beneficial traits (characteristics), such as above-normal eyesight, strong arms, or coloring helpful for camouflage. These individuals are most likely to survive. When they mate, they pass on these characteristics to their offspring. But if a species' surroundings change, its most beneficial traits may change in response. An example of this occurred in 1800s England when pollution from factories turned buildings and trees dark with soot.

As pollution killed the lichens and darkened the trees on which peppered moths had once hidden from predators, the moths became more vulnerable. Their whiter color (previously a beneficial trait) now made them easier for predators to spot. Darker varieties of peppered moths flourished, though. Another example of natural selection is what happened over many generations when female peacocks selected mates that had large, bright-colored feathers. The result was that the male peacocks with duller tails did not reproduce as often and became rare. Such cases point to the idea that individuals with newly beneficial characteristics will tend to live longer and produce more offspring, and over time, the species will look different from how it began.

Darwin used evidence from such species as barnacles, finches, and pigeons to support his natural selection theory. In the fifth edition of *On the Origin of Species*, published in 1869, he used the phrase "survival of the fittest" as part of his explanation of natural selection. He also gathered evidence that all humans originated in

Concepts of evolution in Darwin's On the Origin of Species *became fundamental to the life sciences.*

DID YOU KNOW? In 1996, scientists in Scotland cloned, or made an exact copy of, the first mammal from an adult cell: a sheep named Dolly.

SURVIVAL OF THE FITTEST

Africa in his book *The Descent of Man*. Modern genetic tests have supported many of Darwin's deductions. Despite all the scientific evidence, Darwin's theories were controversial in his time and remain so in some parts of the world today.

After Darwin published his beliefs about human variation and evolution in *The Descent of Man*, it became the foundation for modern biological anthropology. This scientific field centers on researching the varieties of human beings, extinct ancestors, monkeys, and apes—studying their behavior and living processes. Looking at bones, language variations, artifacts, and man-made intellectual objects, biological anthropologists study human evolution.

"Ecology" was not a word when Darwin's journal was first published, but his research also launched that field. Rooted in observation, ecology is the study of relationships between living things and their environment, including if and how humans are still evolving. Some ecologists are studying the sickle hemoglobin **gene** mutation that is most prevalent in people from Africa, where malaria, a parasitic disease that can cause death, is widespread. People who have this gene mutation are resistant to malaria. But the mutation can cause other health problems, such as blood disorders and organ damage. Scientists hypothesize that the constant exposure of malaria caused the mutation to appear many hundreds of generations ago, and because those people survived the disease, they passed on the beneficial mutation to their offspring.

Darwin understood that natural selection was how living things evolved, but he did not know exactly how traits were passed to the next generation. It was Austrian monk Gregor Mendel who discovered this process in 1854. His work supplied the roots for modern genetics. Before Mendel began his plant studies, it was thought that offspring, no matter the species, were a thinned-out mix of traits

23

from both parents. It was also believed that any hybrid (a result of combining two different species) would return to its original form over the course of a few generations. Hybrids were not thought capable of making new forms, either. Part of the reason for this misinformation was because early scientific studies on such topics were not sufficiently lengthy or extensive in scope. But Mendel studied thousands of different plants using the scientific method. He was the first to analyze and hypothesize heredity using statistics, and he kept extensive records over the span of eight years.

Choosing peas for his experiments because of their varieties and ability to quickly reproduce, Mendel cross-fertilized different colors, shapes, and sizes. After analyzing his results, he came up with two heredity laws: the Law of Segregation and the Law of Independent Assortment. The Law of Segregation states that there are both **recessive** and **dominant** traits passed from a parent to an offspring. Traits passed to offspring could come from either parent, and just because the offspring received one characteristic did not mean he or she would get all that parent's genetics, thus the Law of Independent Assortment. Mendel theorized that all living things followed his two laws. Mendel discovered that the way recessive and dominant traits were passed down from parents to offspring followed the mathematical rules of statistics as well. Mendel's laws are now considered foundational to biology.

Mendel (below) recorded such traits as the height of the pea plant (opposite) and the texture of its seeds.

TRY IT OUT! Grow your own mold: Wipe a Q-tip on the floor and rub it on a slice of bread. Drip some water on the bread before placing it in a sealed bag inside an empty milk carton. Keep it away from food and pets! Check your bread each day for fungus.

WHERE DO WE LOOK FOR LIFE? CHAPTER 3

IN THE FIELD

STUDYING BIOLOGY OFTEN TAKES SCIENTISTS AROUND the world. Now a world-famous primatologist, Jane Goodall's interest in animal behavior began in childhood. Goodall spent hours observing birds and other animals and learned that animals have emotions and personalities. She took notes and made sketches of the animals she observed. She also read zoology books. It was one of her dreams to go to Africa and study the animals there.

After receiving her college certificate in 1952, she met archaeologist and paleoanthropologist Dr. Louis Leakey. He hired Goodall as a secretary and allowed her to help with his anthropological dig at Africa's Olduvai Gorge. The Gorge proved to be filled with fossilized remains from ancestors of early humans. Goodall was then sent to Lolui Island in Lake Victoria to study the vervet monkey. "The short study taught me a good deal about such things as note-taking in the field, the sorts of clothes to wear, the

Jane Goodall's research highlighted the humanlike behavior of chimpanzees.

IN THE FIELD

movements a wild monkey will tolerate in a human observer, and those it will not," she said later.

Not long after, Dr. Leakey sent Goodall on a long-term study of the chimpanzees in Gombe, Africa. At the time, not much was known about the chimpanzees, and Leakey believed the study would help provide more evolutionary data. Once in Gombe, Goodall's first attempts to study the creatures ended in failure. As soon as she got within 500 yards (457 m), the chimpanzees scurried away. But Goodall was patient, watching the chimpanzees from far away instead. In time, the chimpanzees allowed her closer.

As she studied the chimpanzees and made notes, Goodall named them after traits she had observed. She called one David Greybeard. In November 1960, she watched David Greybeard make a tool by stripping leaves from a small branch and using it to retrieve termites from inside a mound. This discovery surprised many scientists because it had been widely believed that only humans were clever enough to make tools. Goodall also discovered that chimpanzees were not vegetarians.

Different chimp communities have been found to make tools related to activities such as feeding and fighting.

In 1966, Goodall earned a PhD in ethology, or animal behavior, from the University of Oxford. As she continued researching the Gombe chimpanzees, she was surprised to discover that the animals also had a violent side. She witnessed female chimpanzees yanking infants away from their mothers and devouring them. Dr. Goodall also recorded a "four-year war" between two different chimpanzee groups. Chimpanzees from the Kasakela clan murdered those from the Kahama group until they were eventually all killed.

But Dr. Goodall also found proof of similarities between chimpanzees and humans, such as the chimpanzees' abilities to be affectionate and empathetic. In one case, Goodall and her staff observed

29

SCIENTIST IN THE SPOTLIGHT
RACHEL CARSON

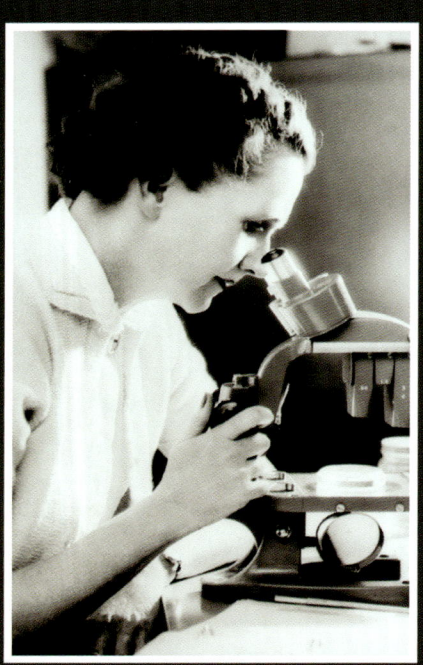

Pennsylvanian Rachel Carson was born in 1907 and explored the local streams and woods as a child. At the Pennsylvania College for Women, she decided to major in English because she liked writing. However, she later changed her degree to biology. In 1943, Carson became an aquatic biologist for the United States Fish and Wildlife Service (then known as the Bureau of Fisheries). She became concerned about the use of pesticides, especially DDT and its effects on fish. Her 1962 book, *Silent Spring*, explained how DDT harms wildlife such as birds and fish, and urged safer alternatives. The book was a controversial success and is often credited with launching the environmentalism movement of the 1960s. Carson testified before a Congressional committee on the effects of DDT, and the pesticide was banned in the U.S. in 1972. The Fish and Wildlife Service later named a part of Maine's coast in honor of Carson, who died in 1964 from cancer.

IN THE FIELD

an orphaned chimpanzee they named Mel. The baby was taken in by a three-year-old male chimpanzee teenager called Spindle. At night, Spindle shared his nest and food with Mel. Spindle let the young chimpanzee ride on his back and stomach and defended Mel against other chimpanzees.

Every year, Dr. Goodall traveled back to Gombe to spend time with the wild chimpanzees as part of the world's longest ongoing wildlife research study. She also encouraged people to take care of other living things. As she wrote in her book *My Life with Chimpanzees*, "... you, as an individual, have a role to play and can make a difference. You get to choose: Do you want to use your life to try to make the world a better place for humans and animals and the environment? Or not?"

In other parts of the world, scientists are working to study and protect wildlife of the oceans. Dr. Ryan Johnson is a marine biologist who has spent his life researching great white sharks. Growing up in New Zealand, he became an accomplished sailor and developed a passion for the ocean. He moved to South Africa to finish his doctorate and conduct research on the great white. In an interview with the *Washington Post*, he said, "The first time you jump in the water with a great white shark without a cage, you never forget that moment, because you can see every gill flutter and movement of its tail.... I remember that first time I hopped in the water like yesterday."

Ichthyology, the study of fish, prompts some researchers to come face-to-face with great white sharks.

On July 24, 2001, Johnson and his team successfully attached a satellite transmitter to an 11-foot (3.4 m) great white female they named Nicole. They have been able to track her migrations back and forth between South Africa and Australia. The project has also

DID YOU KNOW? Australia is home to dangerous animals such as the box jellyfish, crocodile (left), and blue-ringed octopus (above), which has enough poison to kill 23 humans!

allowed them to document shark behavior previously unknown, such as how the great white hunts for seals at night.

From 2001 to 2005, Johnson researched and conducted experiments on the effects of chumming—luring animals with chunks of fish and blood—and the tourist activity of cage diving with great white sharks. At the end of the study, he concluded that "the frequent feeding of these sharks may have begun to alter their behavior." He added that the study possibly showed that the sharks "were associating the cage diving vessel as a reliable source of food." His research did not find a link between such behavior changes and an increase in shark attacks on humans, though. (Contrary to popular belief, the odds of a shark attacking someone are 1 in 11.5 million.)

With the rapid decline of species around the world, conservation biology became a new branch of biology in the 1970s and is referred to as the "discipline with a deadline." As Johnson has written on his website, "The destructive nature of humans towards sharks and marine life is an ever-present burden that is tough to witness. But with more and more people dedicating their lives and efforts to shark conservation, ecotourism, and getting protective legislation in place, I am hopeful that sharks, and the oceans, may one day bite back!"

Botanists at the Daintree Discovery Centre in Australia are employing ecotourism to educate people about the rainforest. They take people from the public into one of the oldest rainforests on Earth, the Daintree, to teach them why the rainforest is important, what plants are in danger of extinction, and what the researchers are studying. Scientists such as Dr. Mike Liddell are monitoring carbon changes and the rainforest's water balance.

Encompassing 463 square miles (1,199 sq km), Daintree is the largest rainforest in Australia. Fossil study has shown that many of its species, such as the idiot fruit, have not changed for more than

IN THE FIELD

100 million years. This makes botanists think the idiot fruit and other ancient plant species found in the rainforest hold many answers to their questions about where all plants came from.

Ecotourism is catching on as more people around the world start to practice environmental responsibility. Conservationist Dr. Baba Dioum once said, "In the end, we will conserve only what we love, we will love only what we understand, and we will understand only what we are taught." Biologists continue to educate others on topics such as recycling, safe waste disposal, and renewable energy so that everyone can do his or her part to help conserve Earth's living things.

Conservationists hope ecotourism and exploration will encourage more people to appreciate nature.

TRY IT OUT! Write the letters "E" and "A" six inches (15.2 cm) apart on a blank paper. Close your left eye. With your right eye about 18 inches (45.7 cm) away, focus on the "E." Slowly move your head forward. The place where "A" disappears is your blind spot!

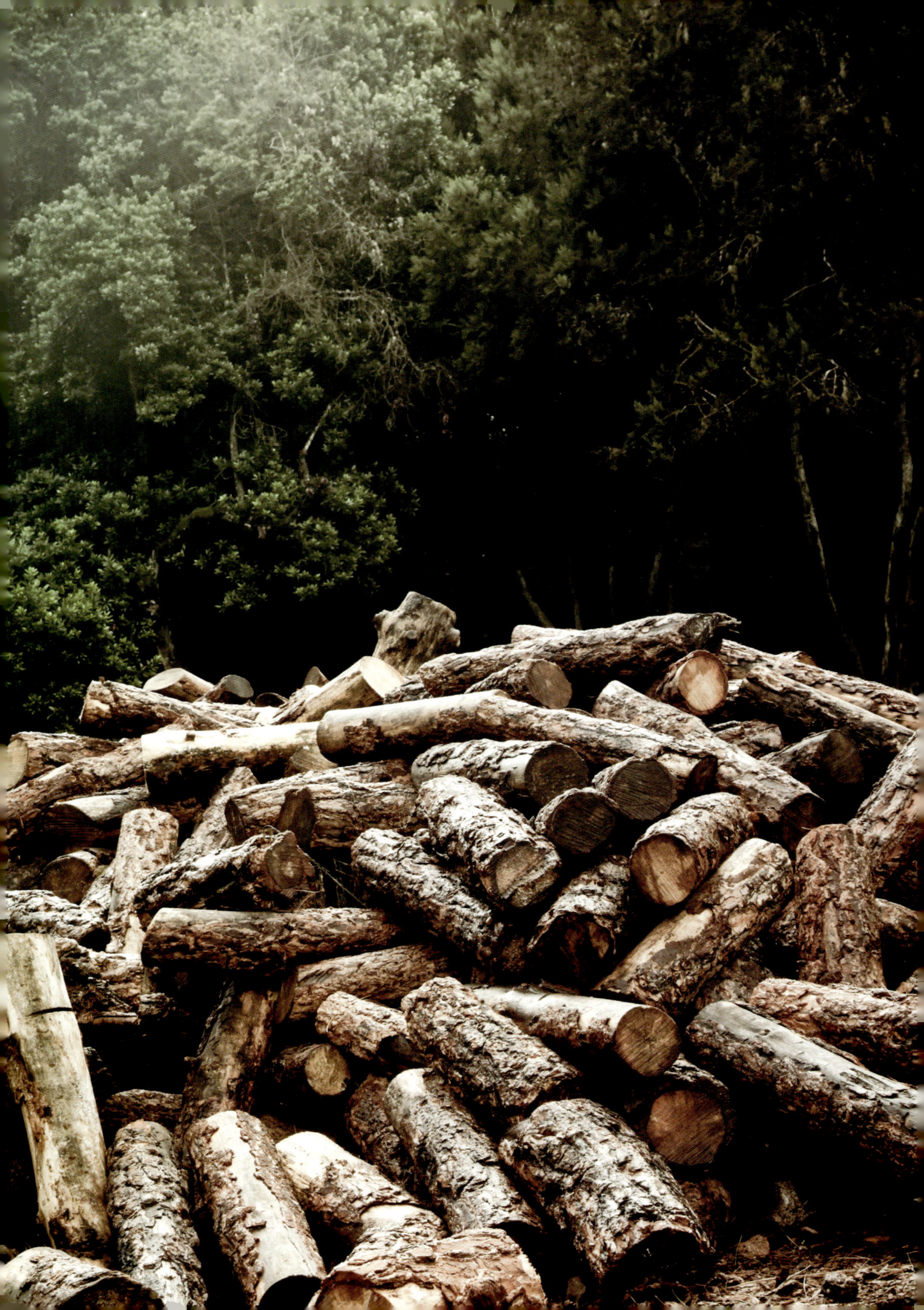

WHERE DO WE LOOK FOR LIFE? CHAPTER 4

YEAR OF THE HORSE

IN 2050, IT WILL BE THE YEAR OF THE HORSE ON THE Chinese calendar, and some life scientists fear that, by then, our planet will be facing a mass extinction event. Some scientists claim it has already begun and we will face by midcentury an even greater amount of species die-off than what occurred 65 million years ago—when dinosaurs disappeared.

Many scientists point to human activity as the reason behind this future extinction crisis. Every second, a rainforest the size of a football field is cut down. At this rate, it is estimated we are losing 5 to 10 percent of rainforest species per decade. The destruction has widespread effects. Rainforests are home to more than 50 percent of Earth's plants and animals and help supply the world's drinking water. Upwards of 2,000 tropical plants have been identified as being helpful in treating cancer—and scientists believe many more remain to be discovered.

Scientists say humans also contribute to species extinction by the accidental

Large-scale deforestation often occurs to make room for farmland or urban areas.

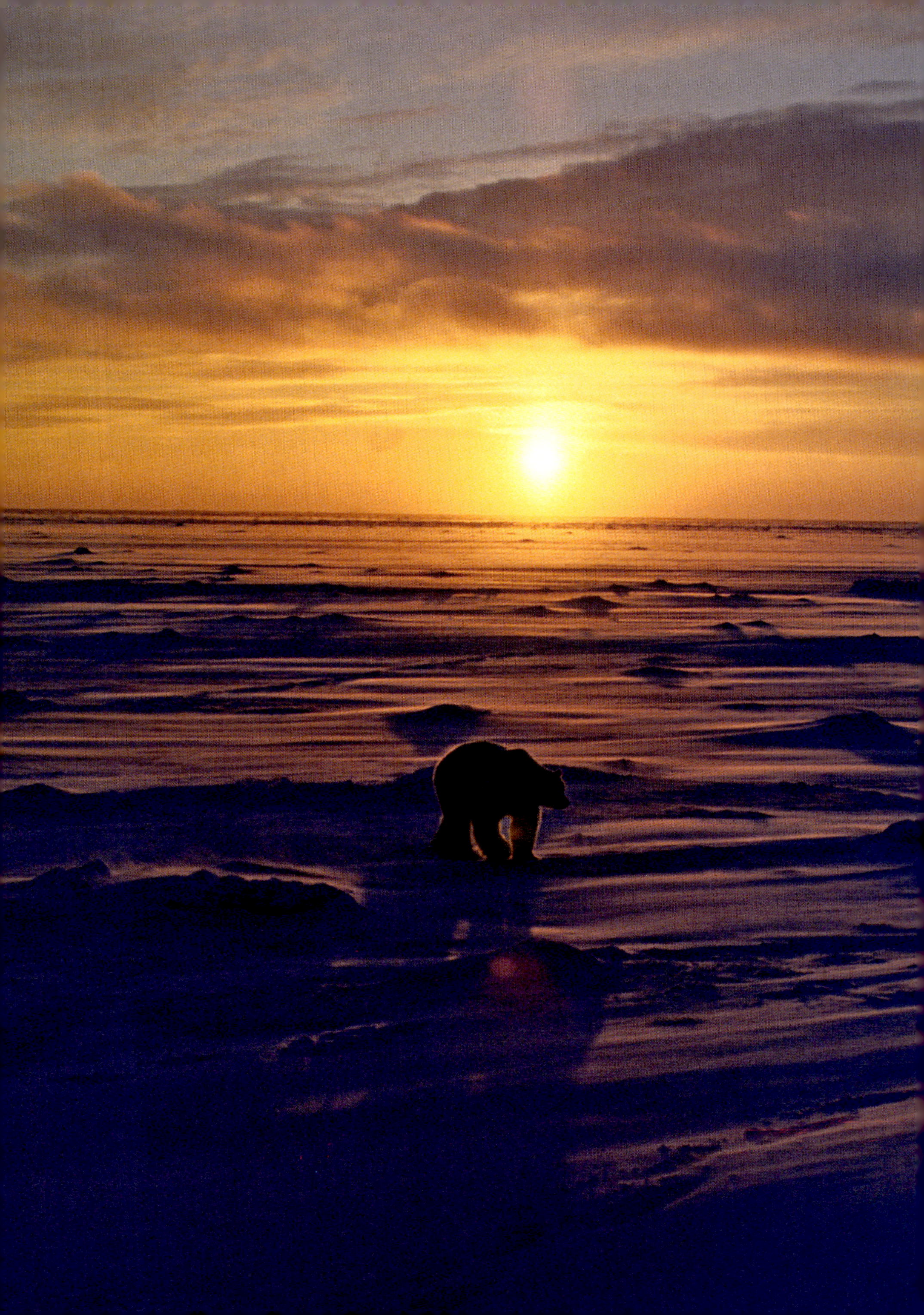

or intentional introduction of nonnative species. An example is on the island of Guam in the Pacific Ocean. Most of the island's native birds are extinct or nearly so because of nonnative brown tree snakes. The snakes likely arrived by boat in the 1950s. Because they had no real predators on the island, their population exploded, reaching an estimated 2 million as they fed off the island's native birds. In late 2013, scientists attempted to curb the overpopulation by airlifting into the jungles dead mice filled with the drug acetaminophen, which is toxic to reptiles. (Unlike other snake species, brown tree snakes do not care if their food is alive or dead but will eat it either way.)

Polar bears (opposite) and all living creatures experience the negative effects of human pollutants (below).

Some scientists blame global warming on humans' carbon dioxide emissions from automobile and factory exhaust. They say our activity is warming up the **atmosphere** and oceans at a rate too fast for species to adapt. **Acid rain** is harming plants and animals. In an event called coral bleaching, the heating up of our oceans causes coral reefs to expel their colorful algae and sometimes die, leaving behind their ghost-white skeletons. These types of harmful occurrences have a domino effect. If one species goes extinct, it can lead to the extinction of other species because food chains are disrupted.

Biologists studying polar bears say that more than two-thirds of current populations will be gone by 2050 if global warming continues melting sea ice. As areas of sea ice get smaller, polar bears have an increasingly difficult time hunting seals, their main food source. In July 2013, scientists found a male polar bear that had starved to death on the Arctic islands of Svalbard. The bear had been healthy just months before. Researchers blamed the death on a lack of sea ice. One scientific study in the 2013 *Proceedings of the National Academy of Sciences* estimates that the Arctic might

SCIENTIST IN THE SPOTLIGHT

ELIZABETH BLACKWELL

Born in England in 1821, Elizabeth Blackwell moved to the U.S. with her family when she was 11. As an adult, Blackwell became a teacher. After hearing a dying friend say she would have been more comfortable if her doctor had been a woman, though, Blackwell began to think about medicine as a career. Male physicians told her it would be impossible—it would cost more than she could afford, and medical schools did not admit women. Blackwell applied to dozens of schools, anyway. Finally, in 1847, she was admitted to Geneva Medical College in New York. Two years later, Blackwell became the first woman in America to receive a medical degree. Along with younger sister Emily and Dr. Marie Zakrzewska, Blackwell founded the New York Infirmary for Women and Children in 1857. Even after she stopped practicing medicine, she continued to educate and inspire women doctors for the rest of her life.

not have any ice left by 2054, and conservationists are calling for immediate action to curb human activities related to climate change. According to the United Nations Intergovernmental Panel on Climate Change, the faster melt rate is spurred by greenhouse gases that build up in Earth's atmosphere and prevent the release of heat. It is believed the gases already in the atmosphere will continue to warm the earth until about 2050, but cutting emissions now might help the species most at risk.

Meanwhile, other scientists are researching the animals that humans eat. More than a billion people around the world depend on fish as their main protein source. But marine biologists say the world's fish populations are in decline because of overfishing. Overfishing happens when water species are harvested faster than they can reproduce. Since 1950, 90 percent of the large predator fish such as sharks, tuna, and swordfish have vanished from the oceans because of overfishing. When the large predators disappear, entire ecosystems become unbalanced. Studying data collected from research vessels between the 1970s and 2005, researchers found that as black reef sharks and other shark species along the coast of North Carolina declined, the scallop industry collapsed. With fewer sharks to feed on cownose rays, the rays' numbers increased, and they ate up all the scallops. Measures to combat overfishing—such as enforcing fishing laws and protecting spawning grounds and coral reefs—can reverse the damage already done, but they would need to be enacted immediately.

Pollution is another problem life scientists are working on solving to ensure a healthier future for living things. Plastic waste makes up an estimated 80 percent of marine litter. Turtles and other animals

Studies suggest that U.S. great white shark populations have slowly rebounded after years of decline.

DID YOU KNOW? A human eats about 2 pounds (0.9 kg) of food a day, a wolf consumes 20 pounds (9.1 kg), and a killer whale feasts on half a ton (454 kg).

often mistake plastic for food. An estimated 44 percent of water mammals have plastic in their guts, which can cause them to starve. Animals are also sensitive to the noise pollution we make. In some cases, humans have used mid-frequency sonar, and mass whale beaching has occurred soon after. With the evidence mounting, sonar exercises may have to be limited to particular areas during certain times of the year.

Besides industrial and pharmaceutical pollution, animals must also contend with eutrophication pollution. This is when chemical runoff from land fertilizers and farm waste is absorbed into the oceans and can destroy coastal ecosystems. Life scientists say that people have the ability to eliminate all these types of pollution by activities such as recycling and making sure their governments work to regulate pollution.

Scientists are also paying attention to the future of humans. With the human population expected to reach more than 9 billion by 2050, life scientists are working on solutions for how humans can still have their transportation, housing, food, and water needs met while coexisting with all of Earth's other species. Scientists at the Food and Agriculture Organization have stated that food production will have to increase by 70 percent by 2050 to meet the world's needs. Some say we will have to adopt a more vegetarian diet because the animals raised as our protein sources require up to 10 times more water. Rather than using land to feed livestock, we will need it to feed people. Currently, most Western diets include about 20 percent of daily protein from animal foods, and that figure may need to be reduced to 5 percent.

Not only will there be more people, but humans will also live longer because of biological discoveries. In the 1800s, people had a life expectancy of 40 years. In the present day, a man living until age 75

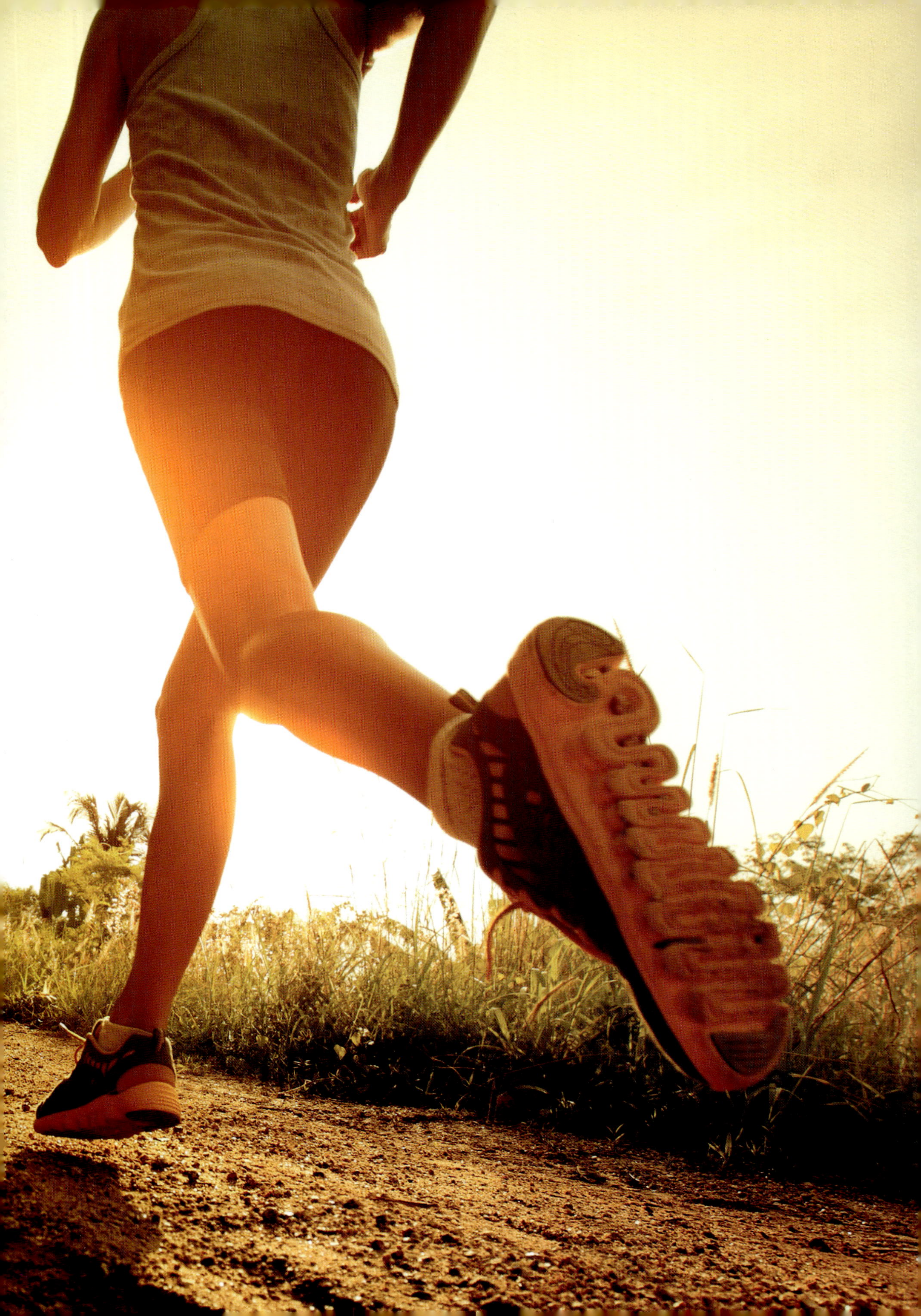

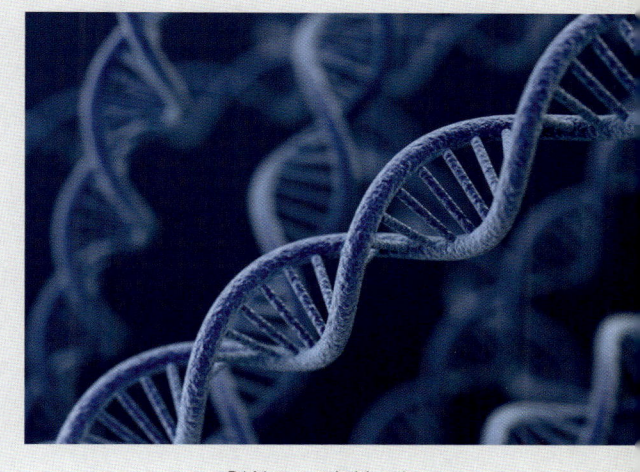

YEAR OF THE HORSE

and a woman living until age 80 is not uncommon. Scientists have determined that the aging process happens at the cellular level. By slowing down degeneration and repairing or replacing faulty materials in our cells and tissues, both quality and longevity of life increase. Medical biologists are working on isolating genes that cause disease and experimenting with using stem cells to repair or replace organs. They are also working to improve medicines to help people better manage illnesses.

DNA research (above) may hold the key to helping people enjoy longer and healthier lives (opposite).

Because of life scientists, we know how to protect our bodies from many types of damage and disease. We can exercise, eat fruits and vegetables, and get vaccinated. Scientists are not trying to have us live longer so that we will be hooked up to a respirator when we are 100 years old. They are aiming for us to feel vibrant, energetic, and healthy, even in old age. Life scientists never stop asking questions, and every day, they make discoveries that can change life as we know it.

TRY IT OUT! Record if the ethylene gas from ripe apples affects unripe fruits and vegetables. Place a ripe apple in a bag with an unripe apple, banana, tomato, avocado, or grapefruit. For comparison, set aside another piece of the fruit to ripen like normal. Which ripens faster?

WHERE DO WE LOOK FOR LIFE? GLOSSARY

acid rain: rain with high sulfur and nitrogen content that harms the environment; often caused by the burning of fossil fuels

adapted: became used to an environment or situation

atmosphere: a mass of gases that surrounds Earth or another planet

culture: in biology, the growth of bacteria or cells in a container

dominant: describing a trait that can appear in offspring even if only one parent contributes the gene

ethical: related to conduct of right and wrong behavior

evolution: the process living things undergo as they develop from one form into another

fossils: the remains of ancient creatures, sometimes found changed into a stony substance or as a mold impressed into rock

gene: the hereditary unit that transfers traits from a parent to a child

heredity: the passing of traits, or genes, from parents to offspring

hypothesized: made an educated guess; suggested an explanation based on a limited amount of evidence

inoculated: implanted weakened or nonliving material in a person or animal to build resistance to a disease

invertebrates: animals without backbones, such as mollusks, arthropods, and others

metamorphosis: an insect's life-cycle process of changing from a young form to an adult form through several transformations

microorganisms: organisms too small to be seen with unaided eyes, requiring a magnifier or microscope for viewing

natural selection: an evolutionary process in which organisms that are better adapted to their environment tend to survive longer and produce more offspring

photosynthesis: a process in which plants, bacteria, and algae use sunlight to convert carbon dioxide and water into food

recessive: describing a trait that appears in offspring only when both parents contribute the gene

scientific method: a step-by-step method of research that includes making observations, forming hypotheses, performing experiments, and analyzing results

superorganisms: individual life forms that function as a larger unit; for example, a beehive

WHERE DO WE LOOK FOR LIFE? SELECTED BIBLIOGRAPHY

Flannery, Tim. *The Weather Makers: How Man Is Changing the Climate and What It Means for Life on Earth*. New York: Grove, 2006.

Geison, Gerald L. *The Private Science of Louis Pasteur*. Princeton, N.J.: Princeton University Press, 1995.

Henig, Robin Marantz. *The Monk in the Garden: The Lost and Found Genius of Gregor Mendel, the Father of Genetics*. Boston: Houghton Mifflin, 2000.

Lear, Linda. *Rachel Carson: Witness for Nature*. New York: Henry Holt, 1997.

Peterson, Dale. *Jane Goodall: The Woman Who Redefined Man*. New York: Mariner Books, 2008.

Ruse, Michael. *Darwin and Design: Does Evolution Have a Purpose?* Cambridge, Mass.: Harvard University Press, 2003.

Steel, Rodney. *Sharks of the World*. New York: Facts on File, 2002.

Wolfe, Art. *Rainforests of the World: Water, Fire, Earth & Air*. Text by Ghillean T. Prance. New York: Crown, 1998.

WHERE DO WE LOOK FOR LIFE? WEBSITES

FLORIDA MUSEUM OF NATURAL HISTORY
https://www.flmnh.ufl.edu/fish/
This website has information on shark research and is the database of the world's human shark attacks.

THE JANE GOODALL INSTITUTE
http://www.janegoodall.org/chimpanzees
Learn all about chimpanzees and biologist Jane Goodall's past and continued work for primates.

Note: Every effort has been made to ensure that the websites listed above are suitable for children, that they have educational value, and that they contain no inappropriate material. However, because of the nature of the Internet, it is impossible to guarantee that these sites will remain active indefinitely or that their contents will not be altered.

WHERE DO WE LOOK FOR LIFE? INDEX

biological anthropology 23

biology 7, 9, 10, 11, 13, 15, 19, 25, 27, 30, 33, 35, 39, 40, 43, 45
    early influences 9, 11
    and medicine 9, 10, 11, 15, 40, 45
    and taxonomy 13, 15
    and world outlook 43, 45

Blackwell, Elizabeth 40
    and New York Infirmary for Women and Children 40

Blackwell, Emily 40

botany 5, 7, 17, 33, 35, 37
    rainforest species 33, 35, 37

Carson, Rachel 30
    DDT research 30

cells 5, 7, 13, 22, 45

conservation efforts 5, 15, 33, 35, 41
    ecotourism 33, 35

Darwin, Charles 15, 17, 19, 21, 23
    coral reef theories 17
    and HMS *Beagle* expedition 17, 19
    natural selection theory 19, 21, 23

Dioum, Baba 35

ecology 23

entomology 11, 13

ethology 29

evolution 15, 21, 23, 29

extinct species 15, 23, 33, 37, 39

fossils 17, 27, 33

genetics 7, 22, 23, 25, 45
    and cloning 22
    mutations 23
    and stem cell research 45

Goodall, Jane 27, 29, 31
    chimpanzee research 29, 31
    and work with Louis Leakey 27, 29

human threats to ecosystems 33, 37, 39, 41, 43
    climate change 39, 41
    introduction of nonnative species 39
    overfishing 41
    pollution 39, 41, 43

immunology 10, 20
    vaccines 10, 20

Jenner, Edward 10

Johnson, Ryan 31, 33
    great white shark research 31, 33

Leeuwenhoek, Anton van 5, 11

Liddell, Mike 33

Linnaeus, Carl 13, 15
    classification system 13, 15

marine biology 5, 31, 33, 41

Mendel, Gregor 23, 25
    heredity laws 25

microbiology 5, 7, 11, 20, 25

microscopes 5, 11, 13

Pasteur, Louis 20
    and pasteurization 20

Schleiden, Matthias 13

Schwann, Theodor 13

scientific method 5, 25

Solander, Daniel 13

Swammerdam, Jan 11, 13
    lymphatic system discoveries 13

Zakrzewska, Marie 40

zoology 7, 17, 27

WHITMAN MIDDLE SCHOOL LIBRARY

WHITMAN MIDDLE SCHOOL LIBRARY